THE CITRUS SEED GROWER'S INDOOR HOW-TO BOOK

THE CITRUS
INDOOR

Also by Hazel Perper
THE AVOCADO PIT GROWER'S INDOOR HOW-TO BOOK

SEED GROWER'S HOW-TO BOOK

By *Hazel Perper*

With Illustrations by Edith Kramer

DODD, MEAD & COMPANY · NEW YORK

_Ø_ For Petesy, with love

Design by Nancy Dale Muldoon
Cover design by Marcia Erickson

ISBN: 0-396-06434-5
Library of Congress Catalog Card Number: 73-179694
Printed in the United States of America

CONTENTS

	PREFACE	7
1	THE CITRUSES AVAILABLE	13
2	THE FRUIT AND ITS RIPENESS	31
3	SIZES AND SHAPES	36
4	SETTING UP FOR PLANTING	42
5	POTTING THE SEEDS	47
6	PRUNING AND CARE	49
7	TRANSPLANTING	57
8	A WORD ON ORIGINS	60

PREFACE

My phone rings. I pick it up and say hello.

"Can I grow two kumquats in one pot?" asks a voice.

The speaker is unknown to me, but I recognize an indoor gardener when I hear one.

"Depends on the size of your pot and on your soil mixture," I promptly answer.

"It's an eight-incher and the soil's left over from a transplanted avocado." Slight pause. "You did write the book, didn't you?"

"Yes, I wrote *The Avocado Pit Grower's Indoor How-to Book*," I admit. "And, please, who is this?"

"Oh, we haven't met, but I read the book."

"And now you're branching out. Tell me, is it correct to assume that we're talking about seeds from a fresh kumquat?"

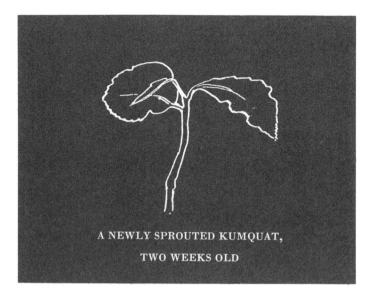

A NEWLY SPROUTED KUMQUAT,
TWO WEEKS OLD

"Of course," says the speaker, cheerfully. "I've got a whole boxful."

"Well, to begin with, it's impossible to say whether your seeds will take."

"Oh, I know *that*. I just wanted to know what *you* think about two kumquats in one smallish pot."

"It's all right for starters. But don't crowd the planting. Place the seeds at least several inches apart."

"How long before they start growing?"

Resisting the impulse to quote Joyce Kilmer, I answer, "It's a matter of time, and some luck. You might add a small handful of sand to the soil mixture, and give some extra plant food as well."

"Hmmmmmmmm." A long, drawn-out sound. I can't tell if it's coming from my kumquat fancier, or if it's a bad connection.

I raise my voice and press on. "Have you noticed that kumquats always seem to have two seeds, never more, never less, always two to a fruit?"

"That's interesting." The voice is barely audible. Then, loud and clear, "But I've got no *sand* in the house!"

"That's easily remedied, isn't it?"

"No, it's not. I'm home from work with a bad cold, and I wanted to use the time to do some gardening."

"In that case forget the sand. Plant the two seeds, and then plant a whole kumquat—a bit dried-out—between the separated seeds."

"How about a wrinkled kumquat? I've got a few of those already. Should I put them on the radiator or maybe in the oven?"

"I wouldn't go that far. Such extreme heat might damage the seeds. Plant them the way they are. Have you tried planting any of the other citruses?"

"No, we drink frozen fruit juice."

"But don't you ever miss the *sight* of the whole fresh fruit?"

"Now that you mention it, maybe that's why I'm enjoying this box of kumquats so much. They're very pretty."

A KUMQUAT, AFTER THREE MONTHS
IN SMALL POT

"Well, then, may I suggest you buy a few oranges and lemons? They're good for colds, and good to look at and to eat. Also, they make charming indoor plants and trees."

"I think there's half a lemon in the icebox. It's a bit old, though."

"Doesn't matter. Just so it has a seed or two."

"Okay. I'll go look. But wait a minute. What about sun, and soil, and cold, and watering, and things like that?"

I fetch a deep sigh. "You're right. A planting plan would be useful."

"Maybe you should write another book."

"Maybe I should."

"Well, I'll call again if my kumquats don't take."

"Well, for both our sakes' I hope they do."

And we say good-by.

And then I decided that the time had come for another book, this one on how to grow citruses indoors. So I wrote it, and here it is.

11

✍ 1

THE CITRUSES
AVAILABLE

In bountiful, beautiful supply citruses are avail-able to everyone everywhere. Most of our fruit is grown in Florida and California, but oranges, lemons, limes, grapefruit, tangerines, and kumquats can be found, bought, eaten, and relished wherever one lives and whenever one likes. The seeds of those fruits, when planted, will become indoor plants and trees that re-

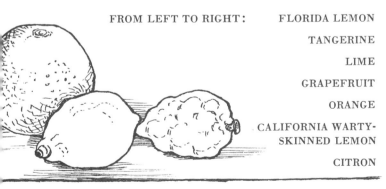

FROM LEFT TO RIGHT: FLORIDA LEMON

TANGERINE

LIME

GRAPEFRUIT

ORANGE

CALIFORNIA WARTY-
SKINNED LEMON

CITRON

13

quire no special care or growing conditions. By the light of one, or two, hundred-watt bulbs I have grown many a healthy citrus plant in indoor sites that were completely sunless.

Readily started indoors, a citrus seedling is a most engaging little plant. Breaking ground, the sprouted seed quickly presents a stand of shiny, green, fragrant leaves supported on a modest but surprisingly sturdy stem—and from the word go that stem seems to declare its every intention of becoming the hardwood trunk of an evergreen tree. Which may well happen. Yet one's plants can be cultivated so they remain at whatever sizes one chooses for them. Started in a small pot, a seedling that has been transferred into successively larger pots will grow progressively taller. Or several seedlings started in a fair-to-middling-sized pot and kept there will remain at relatively low, uniform levels

because their root space has been limited. And since citruses tend to be shrublike in development, they benefit from judicious pruning, which can also help limit their size.

However, while citruses are ideal house plants that will grow lush and full, it may not always be so easy to decide what name to give a particular plant. For the citruses are naturally agreeable to crossing their species lines, and since those lines have also been manually crossed and recrossed, the fruits are varied and many. When a citrus seed germinates and begins to grow, the seedling will retain the characteristics of the parent species—but who is to say which characteristics will be dominant?

In short, gathered together the citruses make for a mixed bag in which familiar fruit with familiar names jostles hybridized fruit with names that may not be so familiar. Orange, lemon, lime, citron, grapefruit, tangerine, and kumquat are household words; not so close to home are pomelo, ugli-fruit, shaddock, pompelmous, limequat, tangelo, and pumplenose. By way of getting a closer look at that lovely motley lot, I began roaming about in citrus country. It was a refreshing junket, and of necessity also a largely bookish one. For a while I felt as if I had wandered into an enormous maze posted with signs that on closer inspection merely pointed to more and more alluring sidepaths. Nevertheless, I managed to gather some fruitful notes on citrus varieties, some of which follow.

CITRON   *Citrus medica*

Because it is traceable to ancient Greece, the citron might well be called the senior citrus. Those early Greeks classified all the fruits known to them as apple, and when the citron reached their shores—for a while it was the only citrus in view—they named it the cedar apple. A suggested reason points to similarities in the shape and foliage of the cedar of Lebanon and the citron tree. Also, a resemblance was supposed to exist between the cedar's greenish-yellow cones and the citron's yellowish-green fruit. Whatever the case,

16

A CITRON

the Greeks dubbed the citron *kedromelon,* meaning
cedar apple, and the Romans picked this up, translat-
ing it as *malum citreum,* which in shortened form,
citreum, was thenceforth applied to the other citruses
in various parts of the widening world. In the eight-
eenth century, Swedish botanist Carolus Linnaeus
chose *citrus* as the generic name for the group, and
although it is a misnomer, and a somewhat odd one,
it has stuck.

As to what a citron may look like, I have never seen
a genuine, live, fresh citrus to match this eighteenth-
century description given by an English lady gardener: "The citron, in Hebrew *ethrog,* is highly valued

for its rind. The fruit is inedible and it measures from five to six inches in length, is oblong in shape with a protuberance at one pointed end, has a rough, pebbled, and adhesive rind with inner portion thick, white, fleshy, and very fragant, and thin outer skin, grey-yellow-green in color." On the other hand, I have planted the seed of big, long, sweet, yellow, warty-skinned California lemons that were obviously closely related to the citron. True, the rinds of my lemon were not especially fragrant, but they were thick enough to be used in the manufacture of the candied citron one finds nestled inside the fruit cakes traditionally eaten at Christmas. This fruit is also ceremoniously presented at the Feast of Tabernacles. Not

A LEMON

surprising then to discover that on the Greek island of Corfu, and in some parts of Israel, to this day the citron is cultivated, albeit in limited quantity. So perhaps one of these days you may find a truly real citron in a local specialty gourmet shop.

LEMON   *C. limon*

The lemon has been proposed as a variety of citron with the lime as possible parent species. It is further said—and as cautiously—that the lemon probably originated somewhere in the East Indies. The tree and fruit migrated to Spain by way of the Crusaders, who between the years 1000 and 1200 reported finding lemons growing all over Palestine.

Today lemon trees grow in almost every tropical and semitropical spot on earth. As to more specific locales, there is no question that not only lemons but all citruses have an affection for seasides. I once knew a citrus enthusiast—Italian by birth—who went so far as to insist that unless a lemon tree was cultivated well within the smell of salt water it would not grow at all.

GRAPEFRUIT   *C. paradisi; C. grandis*

Grapefruit, also known as pomelo—this is *C. paradisi*—is thought to be a satellite of parent species *C. grandis*, whose fruit is known as pummello (variously

A GRAPEFRUIT PLANT, STARTED IN
SMALL POT, AFTER THREE MONTHS

spelled pomello, pumello), also known as shaddock. In other words, our modern grapefruit falls into two species.

Since *C. paradisi* has not been found anywhere in the East Indian archipelago where *C. grandis* is widespread and indigenous, it is said that Jamaica cradled the pomelo. Early botanists described the fruit of the East Indian shaddock as larger than that of the grapefruit, but today in the Caribbean area the two names are interchangeably applied to both fruit and trees, and the fruit tends to be quite uniformly alike. The shaddock acquired its name through the services of a Captain Shaddock. Commanding an English sailing vessel in 1696, the good captain brought the fruit and seeds from the East Indies to Barbados. As to the name *grapefruit*, it seems that early in the nineteenth century on Jamaica a citrus tree was discovered bearing smallish fruit that grew in clusters like grapes. In a similar account the newly discovered fruit was said to taste of the grape.

The grapefruit is the largest, roundest, and plumpest of the citruses, and it is fitting that it enjoy what amounts to a redundancy of names. The earliest name derives from the Latin *pomum*, meaning fruit, especially apple, which probably derived from another Latin word, *opimus*, meaning fat, abundant, stout. From which we can find the makings for pomology, and pompelmous, and pompous, and pumplenose.

I found pumplenose in a nineteenth-century footnote to Marco Polo's thirteenth-century account of his passage through the city of Kamandu, in Persia. The great trader-explorer reports: "On that part of it [the city] which lies nearest to the hills, dates, pomegranates, quinces, and a variety of fruits grow, among which is one called Adam's apple not known in our cool climate." The footnoter adds: "Pomus Adam is a name that has been given to fruit called pumplenose, shaddock, or citrus decumanus of Linnaeus, but here it may be intended for the orange itself, or pomum aurantium, named by the Arabians and Persians 'narani.' (Also spelled 'naranj'.)"

Some forty-odd years into the twentieth century an American botanist reported twenty-three varieties of grapefruit with "normal" colored pulp, and four more with pink or reddish pulp, all of which were propagated in the United States. The grapefruit varieties to choose from are plentiful.

ORANGE *C. cinensis,* also known as sweet, common, or China orange; *C. aurantium,* also known as sour orange; *C. reticulata,* also known as mandarin orange and tangerine

These three species represent a long history of botanical search, research, and discovery that culmi-

AN ORANGE PLANT, STARTED IN
MEDIUM-SIZED POT, AFTER THREE MONTHS

nated in the development of our modern sweet orange. The orange is a beautiful fruit, the best-loved of the citruses—and in itself a symbol of love. An early example of this can be seen on the occasion of Hera's marriage to Zeus, when the distinguished guest Earth presented the bride with the precious gift of a golden orange. The pattern was set—from orange blossoms at weddings to orange juice for breakfasts to generous and inexhaustible quantities of vital Vitamin C for whole families everywhere. The most singularly colored of the citruses is the most popular.

Generally speaking, oranges can be separated into two groups, those from Florida and those from California. The Florida orange has a thin skin, is pale orange in color, and is heavy with juice. The California orange has a thicker skin that peels in chunks, the color is deeply orange, and the fruit tends to be lighter in weight because it holds less juice than the Florida varieties. Within these groups are countless hybrids, both natural and man-made, in numbers far too many to list in these few pages.

TANGERINE   *C. reticulata*

Not so long ago the tangerine was discovered growing wild in Tangiers, hence the name. It is a variety of mandarin orange, which itself is often sold as a tangerine.

24

The Temple orange, named after the grower who developed the fruit, is in effect half sweet orange and half tangerine. It is readily recognizable as a somewhat larger and firmer-skinned version of the smaller and thinner-skinned tangerine. A tangor is said to be half tangerine and half orange, but in my experience it is only a name and not a fruit.

A final word on oranges: my Italian gardener-friend—mentioned earlier—claimed that when he was a small boy in Sorrento, each time the local orange trees came into bloom it was the family custom to bring armsful of fresh-cut branches indoors where they were placed in huge, stone, water-filled jars. This, he explained, was not only for the enjoyment of the foliage and the fragrance of the flowers, but also so that the children might keep close watch on the blossoms as they slowly changed into tiny oranges that turned green as they ripened, and tasted very sweet when eaten. Since fruit will not set under these circumstances, the story is surely apocryphal. But such childhood memories, touched with magic, are not to be denied.

TANGELOS AND UGLI-FRUIT

It has been said that the tangelo and the ugli-fruit —which may not be as familiar as related types of citruses—are the result of crosses between the tangerine and the grapefruit. In the case of the ugli-fruit, however, looking at the fruit one gets the impression that the tangerine side of the family was a good bit more rumple-skinned than usual. These two citruses favor the Caribbean and are not marketed as widely as other varieties.

LIME   *C. aurantifolia*

As distinctive as its cool-green fruit, the lime seems to be the least cold-resistant of the citruses. In the sixteenth century, when Portuguese and Spanish sailor-explorers brought the lime to this continent, the trees took root in the warmer areas of southern Florida, southern California, and Mexico.

Although the fruit is usually to be found in good year-round supply, finding a lime with seeds can be very difficult. Some Mexican limes and the so-called Key limes contain seeds. These two varieties can be found in gourmet shops that specialize in tropical and out-of-season produce. Key limes come from the Florida Keys, and when ripe are yellow rather than green.

However, when a storekeeper tells you that his limes are Persian, it is a practical certainty that the fruit will be thoroughly and completely seedless.

LIMEQUAT

In a gourmet shop I once came upon a box of tiny, round, hard, muddy-green citruses that the shopkeeper

26

assured me were citrons. Sniffing in disbelief, I inhaled the unmistakable fresh scent of lime, and looking at the fruit's odd color, I was smitten with the romantic notion that this might be a long-sought-after limequat. Whereupon I bought three of the unknowns (they were expensive enough to have been transported individually from the far ends of the earth by personal couriers) and found two seeds contained in the surprisingly sweet flesh of each fruit. Three of the planted seeds took root and grew into pretty little plants that we called limequats. But two were cold-sensitive, and only one plant survived its first indoor winter in New York.

Actually, I do not know if the fruit was or was not what I imagine a limequat to be, a literal cross between the kumquat and the lime. I believe, however, that there are limequats—and they are an old-fashioned variety of citrus, a fanciful hybrid cultivated in a long-ago past when the citrus industry was not so completely given over to frozen juice concentrates with the consequent loss of interest in whole fresh fruit.

KUMQUAT *Fortunella japonica; F. margarita*

Until the middle of the nineteenth century kumquats were classified as *Citrus japonica*. Then a British traveler-botanist, Robert Fortune, declared the kum-

A KUMQUAT

quat a separate genus, which was henceforth known as *Fortunella.*

Kumquats may be as small as one inch long or as big as one and a half inches. And the fruit is familiar to us in two varieties: the round Marumi kumquat, *F. japonica,* which is orange-yellow when ripe, and the oval Nagami kumquat, *F. margarita,* which is reddish-orange when ripe. They are said to be widely culti-vated in Japan and China as hedge plants.

28

DWARF CITRUSES

In its natural habitat a citrus seedling needs at least ten years of good growth before it is ready for blossoming and fruiting. And it is a practical certainly that an indoor citrus seedling—no matter what its age —will not produce flowers or fruit. Therefore, to anyone who yearns for an indoor fruit-bearing citrus, I would say, buy a dwarf citrus tree from a florist. As to the kinds of trees you might ask for, the following are the most likely to be found.

Best known, the ponderosa lemon yields large lemons that, on the whole, are inedible; this is a hybrid, probably with the citron. The Meyer lemon, which is a natural dwarf and can be grown from its seeds, combines lemon and orange characteristics. The dwarf otaheite orange supposedly bears fruit all year round, and the Calamondin orange is a handsome small tree with edible sourish fruit. Among limes are the Bearss and a dwarf Persian, which are almost perfectly seedless.

The dwarfs are pretty little trees, but they require specialized care. Instructions for their maintenance warn against temperatures that vary, water in too great quantities, and sun that is too bright. They also want lots of fresh air, and not too much plant food.

I have had small success with store-bought dwarf citruses. I think that is mostly because they demand

29

more time and attention than I care or can find to give. Some years ago I owned a Meyer lemon tree that for quite a while flourished fruitfully. Then, during a holiday season, a host of youthful visitors picked the tree clean. Nevertheless, I managed to rescue and plant two or three seeds, one of which took root and put forth a plant that closely resembled another citrus started about the same time from the seed of an everyday lemon. The latter plant first came into the light of day inside a small glass fishbowl with a glass lid—a vessel intended to function as a terrarium. Soon the glass lid broke and the terrarium plants went their various ways, but the lemon stayed. Today, after five years in a closely confined space and after the performance of one top-pruning operation, the tree is six inches high. It is thick with dainty leaves and is altogether charming. The Meyer lemons, alas, both parent and seedling, died.

🌿 2

THE FRUIT AND
ITS RIPENESS

Citrus fruits fall into three color groups, orange, lemon, and lime. And the spectrum ranges from deep-dusky to vibrant orange for oranges and tangerines, from greenish to pale yellow and pink for grapefruit and kumquats, to clear lemon for lemons, to vivid green for limes. But the ripeness of the fruit and the intensity of its color are not necessarily related.

To begin with, there is only one way the fruit of a citrus can be truly ripened, and that is on the tree. In this country it is illegal to market oranges that are not ripe, which helps explain the excellence of those that reach our tables. At the same time our color expectation for the citruses is so well imprinted that the words and imagery for orange, lemon, lime, tangerine, and such have long since become part of our vocabulary. Which is why we expect oranges to be fully orange-colored, and apparently nothing less will do.

But there is more here than meets the consumer-conditioned eye. In a grove of commercially grown oranges, the fruit that ripens on the trees presents a kaleidoscope of tonalities. There are varieties of orange that turn green as they get riper, and when fully ripe are fully and brightly green. Other varieties start out greenish, turn greener, and then in a switch-over turn orange and remain orange. Still other fruit ripens from pale green to pale orange, in no way resembling the full-blooded orange color we are accustomed to seeing. Climate can be a factor, too. Cool nights and warm days tend to encourage greenness in oranges. Tropical climates where day and night temperatures scarcely vary produce oranges that ripen all-over orange.

To meet our color expectations much of the fruit in our market places has had color artificially applied.

32

So, the colors of citrus fruits are not an indication of the fruit's ripeness, but rather a hint of the fruit's ancestry and source, or the skills of the fruit dyers. I have bought quantities of organically grown oranges that were pale yellow inside and out, had green stem ends, and had the rest of the surface occasionally streaked with greenish and copper-colored pigmentation. The pulp and the juice of this pale yellow fruit were unfailingly sweet, and the seeds germinated with unfailing readiness. I have also bought commercially grown oranges with skins that were uniformly colored from stem to stern, and their pulp and juice varied from pale yellow to pinkish to strong orange, and from sweet to startlingly sour in flavor. Their seeds, too, germinated with unfailing readiness.

✍ CHOOSING SEEDS

Finding seeds that will produce the strongest, leafiest, and best indoor citrus plants is a matter for experimentation. After all, successful gardening is not something as unknown or mystical as the acquisition of a green thumb. In the beginning it is as simple as planting seeds, which in nature is as simple as fruit falling out of a tree.

It has been my experience that seeds removed from

33

fresh fruit freshly cut open take readily. Seeds from fruit kept on hand intact until it begins to soften and turn a bit squishy may germinate sooner. And seeds from oversoft and downright rotten fruit may take soonest. I have never had any luck with fruit so old that it dried to the point where the seeds were desiccated. But since there are always plenty of citrus seeds available I have not done much in the way of working with such seeds.

Occasionally one finds a citrus seed that has burst through its shell and thrust out a slender, hopeful shoot. I always make a point of planting such a seed, not so much because I know it will take root and grow, as because it seems so eager to prove itself.

Finally, it must be said that some citrus seeds produce plants that are stronger, bigger, and prettier than others, but one never really knows which seed may prove to be that kind of a winner.

✐ A WORD ON SEEDLESS CITRUSES

Because of brilliant work done in the field of citrus pomology, today there may be fewer seeds in citruses grown commercially. At the same time, very few citruses are completely seedless. Navel oranges, and one or two varieties of limes, tend toward true

34

seedlessness. But most of the fruit we consume comes to us supplied with seeds, sometimes more and sometimes less. Even so-called seedless fruit is usually not entirely so. Commercial growers market as seedless fruits that contain up to half a dozen seeds.

✍ 3

SIZES AND SHAPES

Citrus plants and trees vary in size in relation to the sizes of expected fruits, yet from species to species the contour of the trees and the shapes of the leaves are closely related. The shapes of leaves grown early-on differ from the shapes of leaves that make later appearances. For example, grapefruit leaves with winged petioles may emerge after the plant has become well-established and well-leafed with singly-formed leaves.

Generally speaking, it can be said that in nature grapefruit trees grow from 15 feet up to 25 feet. The leaves, which taper at both ends (lanceolate), grow dense, dark, and glossy green and are said to be smaller than the leaves of the pummello. The grapefruit is the tallest and fastest-growing indoor citrus and a good tree to grow for relatively quick results.

Orange trees, including tangerines and their kin,

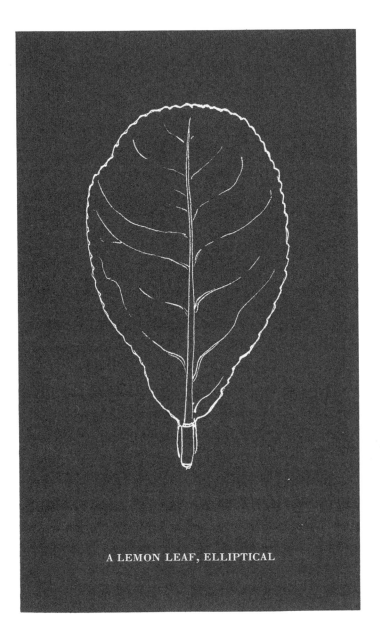

A LEMON LEAF, ELLIPTICAL

TWO VIEWS OF A
CITRUS LEAF WITH
WINGED PETIOLE

grow from 8 feet up to 25 feet, with a natural head of deep green foliage that may be rounded by pruning. Sometimes leaves are blunter at the tips and broader at the center than those of the grapefruit. These trees, too, grow fairly quickly indoors.

Lemon trees grow from 8 feet up to 20 feet. Leaves are paler green than those of the oranges and may be somewhat more scattered about on the tree.

Lime trees grow from 8 feet up to 15 feet and may have a shrubbier appearance than the other citruses. The thick foliage may be made up of smaller leaves, and leaf color may be paler green than on the grapefruit and orange.

Kumquat trees grow from 8 feet up to 12 feet, and their smooth green leaves grow thick and fast.

The above tree sizes and leaf characteristics represent average expectations. It is notable, however, that many citruses in their natural habitats have grown to great heights and ripe old ages. The citrus literature abounds with reports of famous scions, favorite native sons as it were, that have reached up to 60 feet in height, and spread some 30-odd feet in width. Of more modest proportions was a splendid and venerable tree dated 1421 that flourished in the original orangery in the garden at Versailles. It was still alive in 1894.

When cultivated in indoor gardens, citrus seedlings tend toward contours that are tall, slim, and elegant.

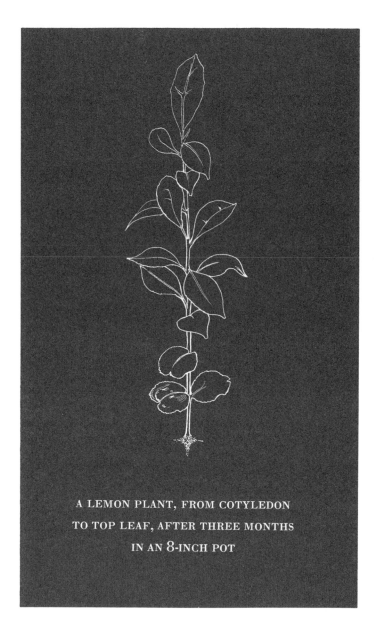

A LEMON PLANT, FROM COTYLEDON
TO TOP LEAF, AFTER THREE MONTHS
IN AN 8-INCH POT

A small plant may be quite conventional in its early shape, and then, as branches begin to appear, these take off and out at often unexpected angles, suggesting the tree's Oriental heritage. This angularity is enhanced by the subsequent offering of a goodly crop of thorns. Characteristic of citrus seedlings, the thorns emerge green, soft, slight and harmlessly pliant. In no time at all they become ferociously sharp and unyieldingly tough. They can be clipped off if you insist, but I prefer to leave them on. Another characteristic is the hard bark that soon overlays the green stem, which, as it becomes a true trunk, will be streaked with long orange-lemon-colored lines that become darker as the tree ages.

41

*4

SETTING UP
FOR PLANTING

You'll need: 1. Flower pots, at least two.
2. Dishes to go under the pots to hold water overflow.
3. Chunks of broken pottery to be used as crocking
material at the bottom of pots. 4. Soil. 5. Plant food.
6. A spray bottle.

*✍ POTS

Pots are important. They should have good
drainage holes. The best ones are the unpainted, in-
expensive brownish-red terra-cotta pots that sweat.

For planting purposes, avoid so-called decorator
pots without drainage holes. If you have a handsome
nondrainable pot that you are particularly fond of,

42

use it as the outer vessel for a terra-cotta pot. But be cautious when watering. The water at the dark bottom of a secondary pot will take a long time to evaporate, and citruses do not do well with their roots trapped in standing water.

There is an indoor-gardening school that extols the virtues of seedlings started in tin cans with holes punched in their bottoms, and there is yet another that presses for the use of paper cups. Obviously both methods can be used. But in my opinion they give unattractive housing to plants, which is a matter of personal taste.

✍ POT SIZES

Most terra-cotta pots are measured by the diameter across the pot's open top, which is more or less the same as the pot's height. To put it another way, a pot six inches across at the top will stand six inches high.

The sizes you need will depend on your planting plans and ambitions. It is well to remember that a citrus plant's underground root structure will be reflected in the plant's aerial structure. When the plant's root space is expanded by a larger pot, the size of the plant will expand.

With this in mind, there are any number of planting plans to choose from. For starters, half a dozen

2-inch pots planted with a seed each from six different citruses offers a source of seedlings that can later be transplanted into successively larger pots, or directly into a good-sized tub. Or six different citrus seeds can be planted in one 10-inch pot, which will allow faster and higher growth. Don't worry about having too many plants. Nothing is quite so pleasant as making a gift of a nicely established plant to an admiring visitor.

DISHES TO GO UNDER POTS

To match terra-cotta flowerpots, there are terra-cotta plates intended to hold water overflow. The effect may give an overall nicety of texture and color, but the plates sweat and deposit small oozing puddles where they are placed. Much easier to keep clean and dry are glass Pyrex pie-plates, which can be found in several sizes. Or use glazed pottery plates, or any other sort of plate that is waterproof.

CROCKING MATERIAL

A broken flowerpot placed over the drainage hole at the bottom of the pot prevents too-dense packing of the root ball. What is needed is a set of curved shards that will keep the soil loose. For this purpose an expendable flower pot 2 or 3 inches in size will give

44

enough material for an 8- to 10-inch pot. A cracked mug or cup will also serve, and for larger pots more shards are needed. To get the crockery broken into not-too-small pieces, fold a newspaper over the flower-pot or mug, and with a hammer smash it into good-sized chunks. Avoid pebbles too small to serve the purpose.

✍ SOIL

Citruses flourish in good, rich soil. Humus-enriched mixtures can be bought in garden-supply stores, but be certain the soil has been sterilized. To this mixture add two cupfuls of sand for a 10-inch pot. Do not use sand that is not fresh and clean. When in doubt, boil the sand in water for a few minutes; then, through cheesecloth, strain off the water.

✍ PLANT FOOD

Commercial plant foods with balanced proportions of nitrogen, phosphorus, potassium, and other needed materials can be found in garden-supply stores, and come packaged in tablet, powder, and liquid forms. It is safer to give too little rather than too much. Follow instructions on the label. On the whole it is difficult to underfeed citruses, provided, of course, food is given on a regular basis, about once in every six weeks.

✐ SPRAY BOTTLE

For the refreshment of the plant's leaves an occasional misting is desirable. It is not necessary to buy a special bottle for this purpose. Use a bottle of the kind once filled with window-cleaning fluid, thoroughly clean it, and then fill with clear tepid water.

46

5

POTTING
THE SEEDS

Whatever the size of your pot, the soil should be well-watered before seeds are planted. Fill the pot to the top with soil, and do not pack it tight. Add water until it flows through into the dish at the bottom.

Use water that is tepid, even warmish, but never cold.

Then plant the seed or seeds. A good rule of thumb is to plant a seed twice the depth of its girth. Citrus seeds take quite shallow planting. When planting more than one seed to a pot, give each seed at least a square inch of space of its own.

It is good practice to keep records of the kinds of seeds being planted. This can be done by clearly marking pots planted with grapefruit, oranges, etc. Or, using toothpicks and small strips of paper on which names of planted seeds are written, place individual banners alongside each different seed.

✒ WHERE AND HOW TO KEEP FRESHLY PLANTED POTS

The best spot for pots is the sunniest place available. If your supply of sunlight is limited, place the pot under the light of a hundred-watt bulb at a distance no closer than eighteen inches. A period of three to four hours a day is adequate. Even better, expose pots to several hours of light from a white frosted fluorescent bulb, at the same distance as above.

Soil should be kept uniformly moist while seeds are germinating.

It is impossible to say how long it will take for the seeds to sprout, but three or four weeks is about the right waiting time. When a sprout appears, place the pot out of direct summer sunlight to protect young plants from sunburn.

✿ 6

PRUNING AND CARE

Do not remove the leaves of citruses until the plants are at least six months old. Only then is it advisable to nip off a top leaf or two. Always use a sharp cutting instrument when pruning.

Leaves at the top of a young citrus often grow larger than lower leaves. Should you decide on a formalized shape—in the sense that a hedge can be cut to form—those top leaves are the first to go. You may also remove any straggling leaves or branches that emerge. The citruses are naturally graceful in shape, and I have relished the handsome sight of many an unclipped plant and tree.

While it is true that timidity and pruning do not go hand in hand, neither does the sort of mustache-pruning that results in snipping off one side of the growth and then the other until a mere tuft remains. In some

49

A PLANT AFTER FIRST PRUNING,
AND AFTER SIX MONTHS IN THE POT

quarters it is believed that any kind of cutting back
or pruning is better undertaken in the spring of the
year. I mention this only because I prune citruses any
time of the year I feel they need it, which to my eye
is very seldom.

50

✍ TEMPERATURES

Cold. Anything under 40 degrees Fahrenheit is likely to be harmful. Although some citrus species withstand cold better than others, it is simpler and easier to avoid extreme chill. In winter, guard against icy blasts from open windows and heed the weather forecasters.

Also in winter, when necessary, supplement sunlight with artificial light. Be careful not to place plants too close to light bulbs, which can scorch leaves. And watch out for dehydrated air, which results from too much indoor heat. Light misting of the leaves is especially helpful here.

Heat. Citruses thrive in most summer temperatures. They favor cool night temepratures: that is, at least 50 degrees. Wherever and whenever possible, potted citruses can be moved outdoors and kept there throughout the summer. City-apartment terraces are especially friendly to growing citruses.

✍ A NOTE ON SOIL

When soil becomes packed too tight, which happens after a while, loosen it up. Use a fork and turn the soil to a depth of at least two inches, working carefully to avoid catching or snagging roots. With the soil thus aerated, the plant is better served.

WHERE TO CUT BACK ON MATURING PLANT

_ A NOTE ON WATERING

Should your plants and trees become dehydrated
—that is, if the top of the soil is dried out and cracked
—first moisten the surface and then pour water into
the dish at the bottom. Wait until the water in the dish
is drawn up into the pot, and then repeat—that is,
water the surface, and then pour water into the dish
below the tub. Do this until the plants are soaked, and
don't water again until they are nearly dry.

_ WASHING THE LEAVES

Citrus leaves may be quite glossy and smooth,
or they may be not so glossy and smooth. Whatever
the texture of the leaves displayed by your plant, they
will certainly do better and look better when washed
shining clean. The trunk of the tree and its branches
also benefit from an occasional dampening. So, use
the window-spray bottle referred to earlier, filled with
clean tepid water. To protect the environment from
too-enthusiastic schpritzing, spread newspapers to
absorb excesses. Or you can use a soft wet cloth to
wash leaves one by one very carefully.

_ FALLING OR YELLOWING LEAVES

It is a good idea to recognize that citruses are
trees, and from time to time their leaves yellow and

53

fall to make way for a stout trunk and resilient branches. When leaves fall, leave them on the surface soil; they nourish the plant.

When your plants are young and leaves fall or yellow, check to see that the plant is not getting too much sun. But remember that in their natural environ-

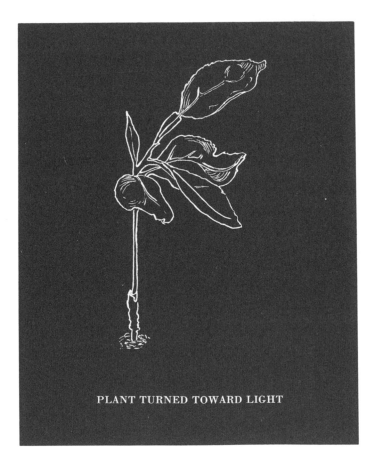

PLANT TURNED TOWARD LIGHT

ment citrus trees flourish even when the weather becomes truly hot and the sun shines strongly from a
clear sky. And because there is something about open
air and natural winds that encourages the production
and setting of blossoms, in the summertime, if possible, try to keep your citruses outdoors.

✍ A NOTE ON ARTIFICIAL LIGHT

Where sunlight is in short supply, place plants
under the light of lamps. Small plants can be moved
within the range of a table lamp. Trees can be lit
with standing lamps that work like spotlights. Make
certain that plants or trees are given light on all sides
by moving either the lamps or the plants.

✍ THE REST PERIOD

The rest period of a citrus is as difficult to de-
termine as that of any other plant. I reason from the
assumption that when the tree's growth seems to be
slacking off, it is because the plant is resting. When
this quietus is then followed by a spurt of fresh activ-
ity, I claim the plant has passed its period of rest.
Whatever the time of year this might take place, leave
the plant to its rest, which is no more than a matter of
a month or so. When the plant begins to grow again,

55

apply fertilizer as encouragement for both plant and gardener.

✍ A NOTE ON INFECTED CITRUS PLANTS

Citrus seedlings grown indoors are wonderfully free from insects, pests, and diseases. I would say that my citruses have been consistently free of infection because they are cultivated in spanking-clean pots filled with sterilized soil and sand. However, should an infected plant appear, I would immediately separate and isolate it from other plants in my indoor garden. Then I might spray it with repeated applications of soapy water. At the same time, in all frankness, it is more likely that I would dump an infected citrus plant. After all, citrus seeds are free from infection when taken from freshly opened fruit, and seeds are always plentiful.

7

TRANSPLANTING

A strong, well-established citrus plant can be moved after it has been about six months in the pot. You can transfer plants from one pot to the next larger size, and so on into increasingly larger vessels. But do not overlook the possibility of starting a citrus tree in its own large tub.

Consider the use of round, octagonal, or squared-off wooden tubs with adequate drainage holes, which can be found in garden-supply stores. I have two octagonal wooden tubs that measure 18 inches across the top and 15 inches in height. These have horizontal brass bands as reinforcements, which may have to be replaced because they eventually weaken and break away. Any sturdy tub or pot with adequate drainage will do for a citrus tree. In case you are thinking of a great round concrete pot, remember that with the added weight of soil, it will be difficult to manipulate this object into the various positions needed to give the tree light and air.

To catch water overflow from a tub, hard rubber plates are sold in garden-supply stores. They come in large sizes and have an edge that allows at least an inch of water. But it is not a good idea to let a tub stand for long in water. Should you feel tempted to grow two plants in a large tub or flowerpot, think twice before doing so. One root system to a tub or pot makes a better environment for the tree than two root systems in probable conflict, although crowding the plants can help in keeping them smaller in size.

From flowerpot to tub, a safe trip is possible if you proceed with caution. Get a good grip on the plant's stem—between the fingers of one hand—and turn the pot upside down. Give the pot a good thump against the edge of a solid surface, and the plant can be decanted without injury. Should you prefer, as I do, to set the potted plant on a piece of spread newspaper, you can smash the pot with a hammer. This is a superior method, for it does no injury to the plant, offers a supply of crocking materials, and can be enjoyable for its own sake.

Now spread crocking material across the bottom of the tub. Pour in a layer of mixed sand and soil to cover the shards—again being certain that the sand

and soil are sterilized. Fill the tub until the space will just accommodate the plant's root system. Then lift the plant into the tub and fill around the edges with soil. The level of soil should be about one inch below the tub's rim. Water the soil freely and give a generous supply of plant food. Then stand by for wondrous developments.

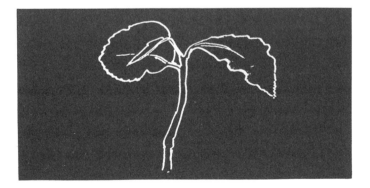

✐ 8

A WORD ON ORIGINS

The citruses are tropical and semitropical by nature, and a long, long time ago they evolved somewhere in southern China, probably in the vicinity of the South China Sea. Then people began to stir about, and as they too evolved they began traveling and the citruses traveled with them. From the Asian mainland the citruses went to Formosa and on up to Japan; they turned south and spread throughout the East Indies and down to Australia; they moved eastward across India, Arabia, and Persia, and settled in parts of North Africa and all over the Mediterranean basin. In the ships of Columbus they came to the New World, in the wake of Pizarro reached Peru.

And in the Garden of the Hesperides there grew a golden tree with golden leaves and boughs bearing

60

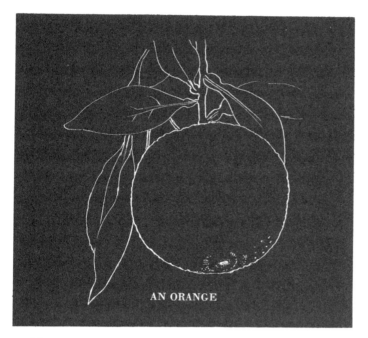

AN ORANGE

golden apples. Claiming those apples, botanists have
called the fruit of a citrus a hesperidium. In Greek
mythology—where so many of our beginnings are
found—the golden tree and golden fruit are the exclu-
sive and highly prized property of the gods of Olym-
pus. Under the zealous care of the seven beauteous
daughters of Atlas, and under the unsleeping eyes of
the fearful serpent Ladon, the golden apples are jeal-
ously guarded against theft by lesser beings. Even the
site of the Garden is a well-kept secret. Rumor has it
on the borders of Ocean, a faraway place where the
sun sets. Yet it is well-known that the sun, held in a

61

golden bowl fashioned by Vulcan, sails across the abode of the Hesperides before it reaches another and still more remote land, where it rises again.

Within that solar passage the golden tree is secure under a beneficent sun, and we are offered a hint of the citruses' affinity with the shores of warm seas. Back on Earth and in sidereal time, change is still in the nature of things. Golden apples and golden myths have yielded to the marvels of hybridization, and the citruses have found countless hospitable landfalls. For a while the cultivation of the trees fell into the appreciative hands of the aristocracy. Sultans, shahs, mandarins, potentates, poets, and artists have cherished the great and noble family of citruses. Kings, queens, princes, minstrels, limners, architects, painters, medicine men, and artisans have celebrated the golden trees and fruit, whether grown in orchards or housed in orangeries as variously structured as the palaces they adjoined. By the time Louis XIV of France had become known as the Sun King, citrus pomology had developed into a fine art. In the middle of the seventeenth century the fabulous gardens at Versailles contair.ed some twelve hundred citrus plants and trees in long roofed-over galleries open at the south side and glassed over on the north side. And with such and other like precursors of greenhouses to come, the citruses moved into the public domain.

62

It is said that in eighteenth-century England, and in those European countries where the climate had proven too cold for their outdoor cultivation, citrus plants and trees were extensively cultivated by indoor gardeners for the sheer beauty of the foliage alone. Grown and kept in pots and tubs—some of which were supported on platforms with wheels to give mobility—the plants were brought out into the warm air of summer, and at the first hint of frost were whisked indoors again.

Finally, and quite lately, commercial citrus groves have been established in countless countries. In southern Florida and California, citrus trees flourish in beautifully maintained plantations that extend as far as the eye can see. In one such grove it seemed to me that the citruses' many attractions have provided them with a set of survival elements. Blossoms so pungently scented that they intoxicate bees and people alike will always be pollinated. Brightly colored fruit that invites instant tasting and rewards with gushes of juice, sweetly bland or bitingly acid, will never lack attentive cultivators. And shining green and graceful foliage that when brushed or crushed between the fingers exudes an evocative fragrance as mouth-watering as freshly opened fruit will always be treasured. Provided, of course, that each and every one of us at once sets out to secure and cherish all living plants, people, creatures, and legends for all time to come.

64